HANDBOOK OF ENEMY AMMUNITION

PAMPHLET No. 3

GERMAN AND ITALIAN AMMUNITION

The Naval & Military Press Ltd

published in association with

ROYAL ARMOURIES

Published by
The Naval & Military Press Ltd
Unit 10 Ridgewood Industrial Park,
Uckfield, East Sussex,
TN22 5QE England
Tel: +44 (0) 1825 749494
Fax: +44 (0) 1825 765701
www.naval-military-press.com

in association with

ROYAL
ARMOURIES

In reprinting in facsimile from the original, any imperfections are inevitably reproduced and the quality may fall short of modern type and cartographic standards.

[*Notified in A.C.Is. 3rd September, 1941.*]

226
Publications
52

> **NOT TO BE PUBLISHED**
>
> The information given in this document is not to be communicated, either directly or indirectly, to the Press or to any person not holding an official position in His Majesty's Service.

HANDBOOK OF ENEMY AMMUNITION

PAMPHLET No. 3

GERMAN AND ITALIAN AMMUNITION

By Command of the Army Council,

THE WAR OFFICE,
3rd September, 1941.

LONDON:
Printed under the authority of HIS MAJESTY'S STATIONERY OFFICE
by Hazell, Watson & Viney, Ltd., London and Aylesbury.

1941

HANDBOOK
OF
ENEMY AMMUNITION

CONTENTS TABLE

German Ammunition

		PAGE
Bomb, Aircraft, H.E., Anti-Personnel, 2 Kg.	Figs. 26, 26a & 27	3
Shell, H.E., 20 m.m., and Fuze, A.Z.1502	Fig. 28	6
Flare, Picket	Fig. 29	9
Bomb, Aircraft, Incendiary/H.E., 1 Kg.	Fig. 30	11
Anti-withdrawal Device, ZUS.40	Fig. 31	12
Shell, H.E., 8 c.m., M.30	Fig. 32	15
Collapsible Container for Incendiary Bombs.	Fig. 33	15

Italian Ammunition

Bombs, Aircraft, H.E., Anti-Personnel (Spezzone).	Fig. 34	18
Fuze for the Destruction of Abandoned Aircraft.	Fig. 35	20
Small Arm, 12·7 m.m., Incendiary Tracer	Fig. 36	21
Small Arm, 12·7 m.m., A.P., Incendiary Tracer.	Fig. 37	22
Small Arm, 7·7 m.m., A.P., Incendiary (Blue Tipped).	Fig. 38	2
Small Arm, 7·7 m.m., A.P., Incendiary (Green Tipped).	Fig. 39	25
Bomb, Aircraft, H.E. (Manzolini), 4·5 Kg.	Figs. 40 & 41	25

RAL: 19453

HANDBOOK
OF
ENEMY AMMUNITION

GERMAN 2-Kg. ANTI-PERSONNEL AIRCRAFT BOMB (" BUTTERFLY BOMB ")   (Figs. 26 and 26a)

This H.E. bomb, fitted with a mechanical fuze which has an alternative time or percussion action, makes use of a hinged outer casing to arm the fuze during its fall. The time action of the fuze is not variable, so that when the time of arming and running exceeds the time of flight the effect of a delay fuze is produced, *i.e.*, a short period will elapse after impact before detonation occurs.

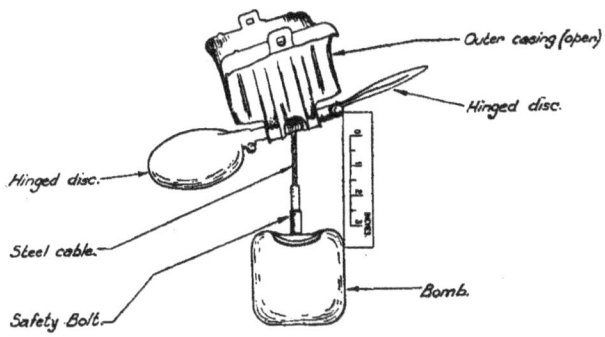

FIG. 26.

The bomb body is a cast-iron cylinder (D) 3 inches in diameter and 3.1 inches long, the average wall thickness being about ⅜ inch. A screw-threaded fuze hole is provided in the side of the body. The interior is coated with a bitumen composition and contains a bursting charge consisting of 7½ oz. of cast T.N.T. topped with the bitumen composition and designed to form an exploder cavity.

The outer casing is of steel and encloses the bomb. The casing consists of two half cylinders connected by hinging at one side and by a securing pin passing through two lugs on the other side. Carried at each end of the hinge bar is a hinged disc. These discs form the end pieces of the cylindrical casing

and are positioned by internal flanges formed at the ends of the half cylinders. The hinges of the discs are inclined in such a way that the discs in the open position are set at a pitch similar to that of an arming vane. Springs are fitted to the discs and half cylinders which cause them to open on their

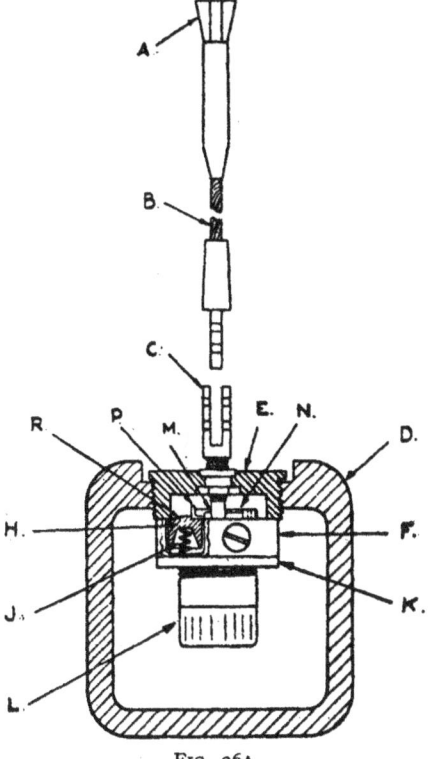

FIG. 26A.

hinges when the safety pin is removed. The outer casing is connected to the head of the safety bolt of the fuze (C) by a short length of steel cable (B). The cable is fitted with a ste head (A) for attachment to the outer casing.

Fuze (Fig. 27)

The fuze consists of three main parts, the cap (E), the body (F) and the base ring (K). The cap and body appear to be made of diecast alloy—probably zinc base—and the base

ring of aluminium alloy. The three portions are bolted together by 3 bolts.

The cap, which is cylindrical, is fixed to the upper side of the body and is screw-threaded for insertion into the bomb.

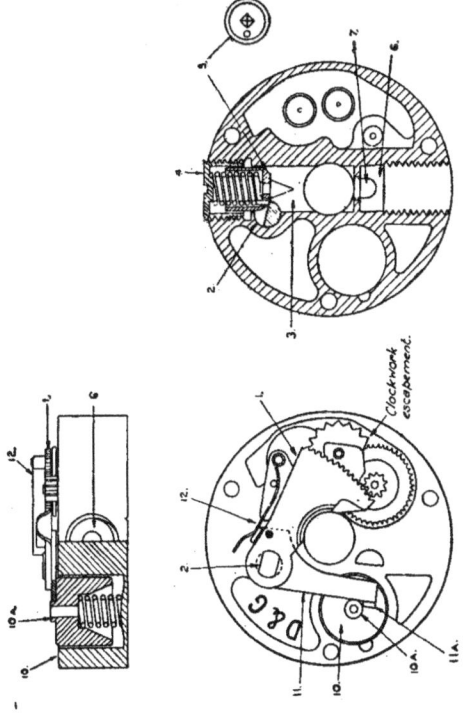

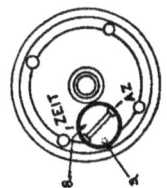

FIG. 27.

On the underside it is recessed to house the mechanical portions fitted on the body and a central hole at the top is provided with a screw-threaded bush to receive the safety bolt. On the upper side of the cap a setting plug (8) is provided by means of which percussion or time action can be arranged. The cap is engraved with two index lines, one marked "AZ" (Aufschlag Zunder, *i.e.*, impact fuze) with which the slot in the setting plug is aligned for percussion action and the other marked "ZEIT" (Time) which is the setting mark for time action. Beneath the setting plug is positioned a small pin (9) which is moved in towards the body when the plug is set to "ZEIT" and withdrawn when set to "AZ."

The safety bolt is connected at its outer end to the connecting cable of the outer casing. The bolt is provided with a double thread for a portion of its length to engage the bush in the cap and is provided with a collar at its inner end which limits the extent to which the bolt can be withdrawn. When screwed home the safety bolt prevents the operation of the clockwork mechanism and masks the detonator (6) from the striker (5).

The body, which is also cylindrical in shape, carries the striker and detonator assemblies in a lateral channel (3). Communicating with this channel is a vertical channel in the centre of the body to receive the safety bolt and a small vertical channel (7) adjacent to the detonator which leads to the gaine. A semi-circular recess is formed in one side of the lateral channel to house the retaining shaft (2). The striker assembly consists of a spiral spring (which also provides the motive power for the clockwork mechanism) held under compression between an outer sleeve screw-threaded for assembly in the body and an inner sleeve which carries the needle. The striker is retained with its spring under compression by the retaining shaft. The portion of the shaft which engages the striker sleeve is semi-circular in cross-section, while the upper end has flats formed on it for the attachment of the time lever (1). The outer end of the lever is in the form of a toothed segment which is enmeshed with the clockwork escapement housed in a recess in the body. A portion of the lever is cut away to accommodate the safety bolt. The lever is also shaped to engage a projection (11A) on the impact arm (11). The impact arm is pivoted on the cylindrical portion of the retaining shaft below the time lever. The strip spring (12) bearing on an extension of the arm tends to make it rotate around the retaining shaft in a clockwise direction. This rotary movement of the arm is prevented by the projection (10A) on a spring loaded detent (10) housed in a recess in the body.

The base ring is screw-threaded internally to receive the gaine, which is of bakelite and encloses a container of similar design to that of the H.E. unit fitted to some of the German 1 Kg. incendiary bombs (*see* Fig. 30).

Action

When the securing pin of the outer casing is withdrawn the hinged half cylinders and end pieces of the casing are forced to the open position by their springs. The half cylinders acting as a drogue in the air stream of the falling bomb causes the connecting cable to tauten and the end pieces, rotating as the result of their pitch, transmit this turning movement to the safety bolt by means of the connecting cable. After three revolutions the safety bolt is unscrewed sufficiently to unmask the striker channel and a further three revolutions removes it from the path of the time lever. The collar on the lower end of the safety bolt prevents the bolt being entirely withdrawn from the fuze cap. The subsequent action of the fuze depends upon the position of the setting plug.

(a) Set to " AZ " (Percussion Action).

On the withdrawal of the safety bolt from contact with the time lever the striker, under the motive power of its spring, imparts a turning movement to the retaining shaft. The time lever, being rigidly attached to the retaining shaft, turns with the shaft in a clockwise direction under the control of the clockwork escapement. The impact arm, not being rigidly fixed to the retaining shaft, is held stationary by the projection on the detent. The rotation of the retaining shaft and time lever continues until the time lever is obstructed by the projection at the end of the impact arm. With the retaining shaft rotated to this position the curved surface of the shaft is almost clear of the striker channel. On impact the detent with its projection sets down, releasing the impact arm, which is then rotated by the action of the strip spring. This movement of the impact arm enables the time lever and retaining shaft to continue the turning movement and free the striker, which is then driven on to the detonator. This action on impact is practically instantaneous.

(b) Set to " ZEIT " (Time Action).

The action of turning the setting plug to " ZEIT " causes the small pin beneath it to move downwards and depress the spring loaded detent, thus removing the projection on the detent from the path of the impact arm. Under the impulse of the strip spring the impact arm is rotated clear of the path of the time lever. The subsequent action is then as described in (a) except that the time lever is not obstructed by the rojection on the impact arm and completes its rotary move-.ient with the retaining shaft. The striker is thus released and driven on to the detonator. The time which elapses between the withdrawal of the safety bolt and the firing of the detonator has been found to vary between two and five seconds with different fuzes.

The flash from the detonator passes to the initiator in the

gaine by the flash channel (7) and brings about the detonation of the bursting charge by means of the P.E.T.N./wax filling of the gaine.

Marking

The outer casing is painted field grey.

GERMAN SHELL, H.E., 20 m.m., AND FUZE, A.Z.150
(Fig. 28)

From information received in later reports, the position of the aluminium supporting ring under the driving band of this shell as shown in Fig. 20 of Pamphlet No. 2 is incorrect. This

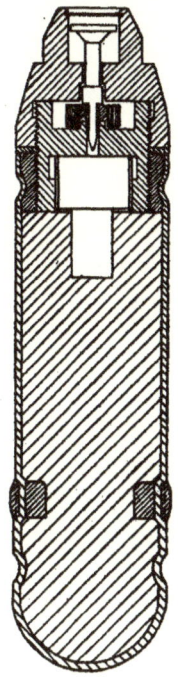

FIG. 28.

figure and the information given on page 2 under the heading "20 m.m. Shell" should be deleted.

The shell body is of drawn steel with thin parallel walls and a hemispherical base. Near the base a cannelure is formed for the attachment of the cartridge case and above this a copper

driving band is fitted in a groove in the body. An aluminium ring is fitted inside the body coincident with and providing support for the driving band. A steel ring with an internal screw-thread for the attachment of the fuze is secured in the head of the shell by four indentations.

The bursting charge consists of approximately 262 grains of penthrite/wax. The filling is designed to form a cavity to receive the magazine portion of the fuze and is covered by a paper washer.

The body of the shell is painted yellow with a black ring near the head.

The fuze, which is inserted with a steel washer between its underside and the head of the shell, is the A.Z.1502. This is a direct action type with its magazine in the form of a brass container fitted to the underside. The container carries a detonator positioned over an intermediary charge of penthrite powder. The needle of the fuze is of the floating type and is held off the detonator by two segments assembled under the enlarged head of the needle. The upper end of the segments and the underside of the enlarged head are shaped to retain the segments in position. The segments are further secured by a length of brass tape coiled around them. Positioned in a recess in the nose of the fuze, above the needle, is the hammer, which consists of a short spindle with an enlarged head. The recess is closed by a metal plate.

Action

On acceleration the needle sets back and holds the segments securely. During flight " creep " action and the protection from air pressure provided by the metal closing disc result in a forward movement of the hammer and needle. When the rate of spin has increased sufficiently the coiling of the brass tape is loosened and the segments thrown clear by centrifugal force. On impact the needle is driven on to the detonator by the hammer and the detonation of the bursting charge is brought about through the penthrite powder charge.

GERMAN PICKET FLARE
(Fig. 29)

The flare, which is probably an army store, is 25.5 inches long, 1¼ inches in diameter, and weighs approximately 2½ lb. The body is a zinc tube fitted at the lower end with wooden spike. The tube is filled with 850 grams (1 lb. 14 oz.) of flare composition, primed at the upper end with 37 grams (1.35 oz.) of loosely stemmed priming. Partly embedded in the priming composition is a blob of match-head composition, the lower part of which is covered with a priming paste. The match head is protected by several discs of crepe paper. Above these is a loose millboard washer half covered with striker paste.

The tube is closed by a zinc cap secured with adhesive tape. The flare is ignited by means of the loose millboard washer and the match head and burns with a white flame.

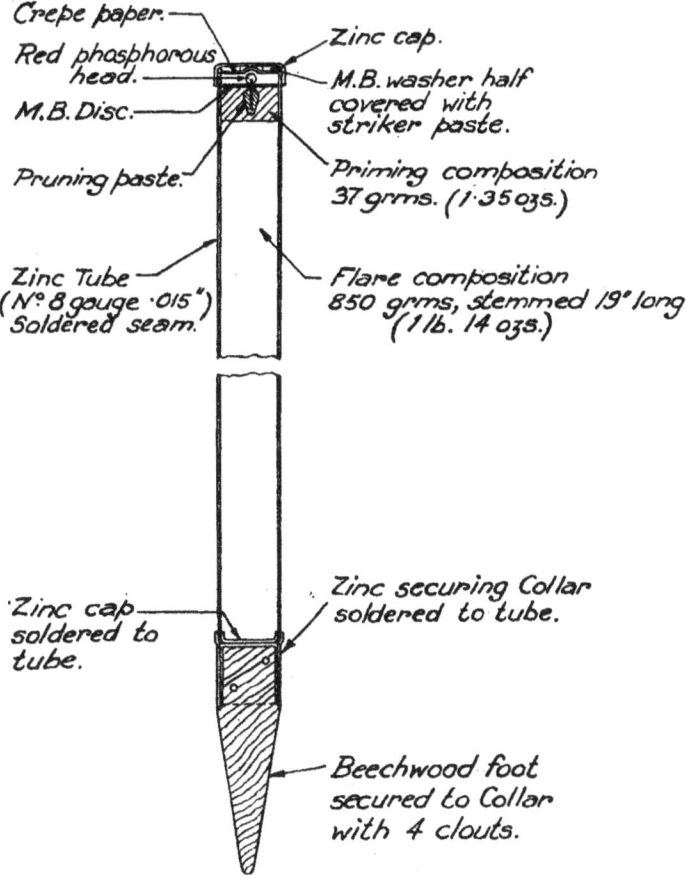

FIG. 29.

Time of burning, 5 mins. 35 secs.
Rate of burning, 17 secs. per inch.
Intensity, candle power—3,200 falling to 850 ; average 1,300.
Efficiency—580 candle secs. per gram.

GERMAN INCENDIARY BOMB, 1 Kg., WITH H.E. BURSTING CHARGE (Fig. 30)

This bomb differs from that described and illustrated in Pamphlet No. 1, Sec. 18, by the inclusion of a small H.E. unit

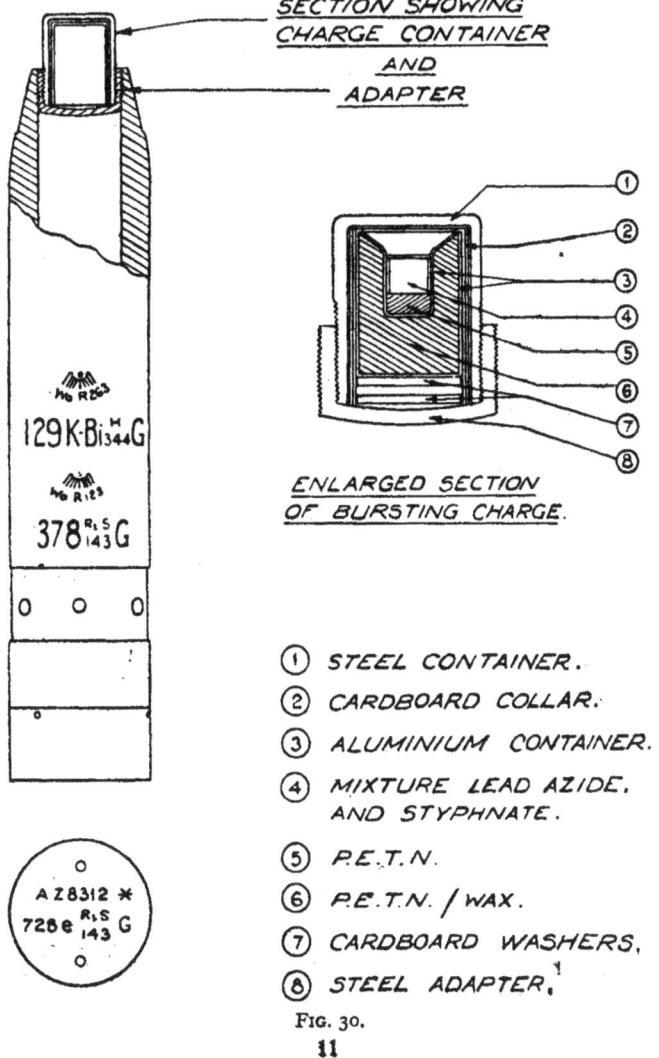

SECTION SHOWING CHARGE CONTAINER AND ADAPTER

ENLARGED SECTION OF BURSTING CHARGE.

① STEEL CONTAINER.
② CARDBOARD COLLAR.
③ ALUMINIUM CONTAINER.
④ MIXTURE LEAD AZIDE. AND STYPHNATE.
⑤ P.E.T.N.
⑥ P.E.T.N. / WAX.
⑦ CARDBOARD WASHERS.
⑧ STEEL ADAPTER.

FIG. 30.

in place of the screwed plug at the tail end of the bomb body.
The H.E. unit consists of a 7·23 gms. bursting charge of P.E.T.N./wax (89·3 per cent. penthrite, 10·7 per cent. wax) in a container of thin sheet aluminium closed at one end. The other end is closed by the aluminium detonator sheath, which is pressed into the explosive charge and secured by turning over the rim of the aluminium container. The detonating composition is a mixture of lead azide and lead styphnate (0·47 gms.) in an aluminium cup pressed into the mouth of the tubular portion of the detonator sheath. Below this, in the bottom part of the detonator sheath is an intermediary of lightly consolidated P.E.T.N. The complete unit is assembled with a cardboard collar and packing discs in a steel capsule approximately 1 inch in diameter, 1·3 inches long and ·08 inch thick. The steel capsule is screwed into a cup-shaped steel adapter which in turn is screwed into the tail end of the bomb body under the cone of the tail unit.

Action

The detonation of the bursting charge is initiated, through the medium of the detonator and P.E.T.N. intermediary, by the burning of the bomb. The time between ignition and detonation is between 1 and 1·5 minutes, by which time the bomb is well alight. The magnesium casing is for the most part fragmented and scattered in a fan-shaped area towards the nose of the bomb over a distance of about 15 feet with occasional projections up to 30 feet. The dispersed fragments are small and burn for only a short time. The nose plug of the bomb is not projected and generally continues to burn, whereas the tail unit, if still in position, would probably be projected rearwards and might produce fragments of a dangerous character.

Marking

Blind bombs of this type have been discovered with the letter " A " stencilled in red on the base. This is probably an identification marking.

GERMAN ANTI-WITHDRAWAL DEVICE, ZUS.40
(Fig. 31)

This device is intended to cause detonation of German H.E. bombs if an attempt is made to withdraw Rheinmetall fuzes from them. The device can be applied to any German Rheinmetall electric fuze fitted with a gaine, but is most likely to be used in conjunction with the long delay fuze El. A.Z. (17).

The drawing indicates the arrangements of the device except that its gaine, which screws into the bottom, is not included.

The needle member (1) is pressed forward by the spring (2)

but is held away from the detonator (5) by a detent (3). The latter is located by the ball (4) below it and maintained in position by the pressure exerted by the needle member (1).

When the gaine of a Rheinmetall fuze is inserted in the top of the device the needle member (1) is pushed outwards

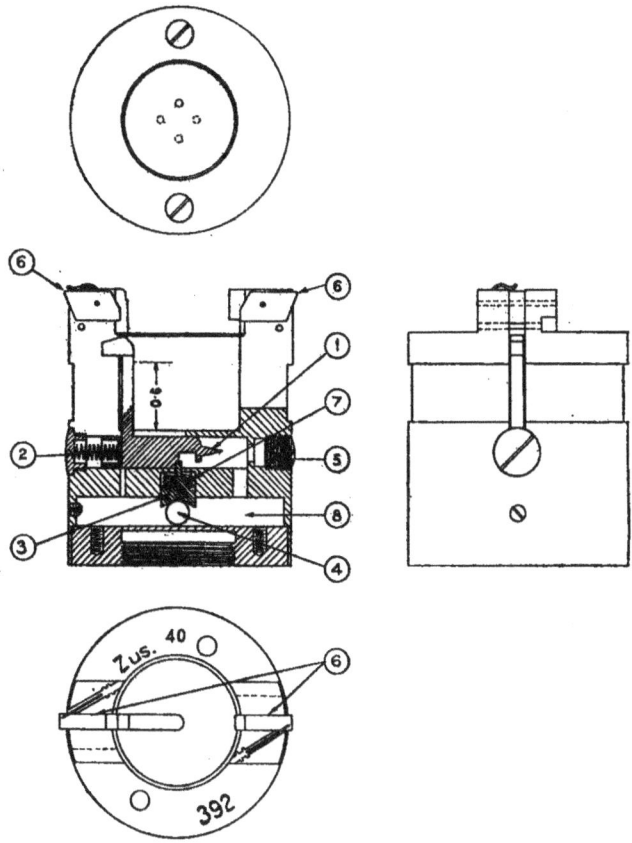

FIG. 31.

slightly, thus removing the frictional constraint from detent (3), which is now located by ball (4) under the action of the detent spring (7).

The device is not armed until the force of impact, transverse

to the fuze, shakes the ball (4) out of the conical recess in the detent (3), which is then pushed clear of the needle (1) by the spring (7) and all three components fall into recess (8).

If an attempt is made to withdraw the fuze from the bomb the device is pulled upwards also owing to the pressure of the needle member (1) on the gaine of the fuze. This movement is prevented, however, by the two knife edges (6) held outwards

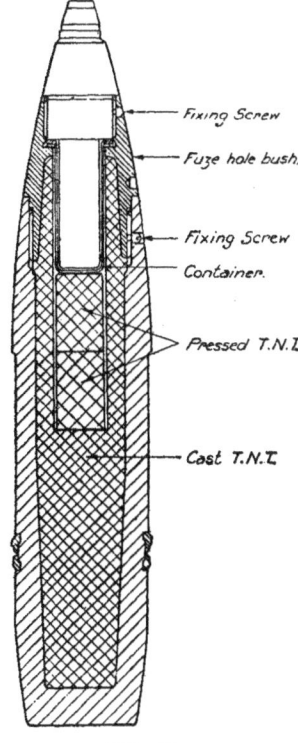

FIG. 32.

by springs which bite into the wall of the exploder tube of the bomb. The gaine of the fuze is therefore withdrawn from the device and after a movement of about 0.6 inch the needle member (1) is clear of the gaine and is forced by the spring (2) into the detonator (5). The flash from the latter passes to the gaine screwed into the bottom of the device (not shown on the drawing) and the bomb is detonated.

GERMAN 8 c.m., H.E. SHELL, M.30
(Fig. 32)

This Czech shell, known as " 8 c.m. A.Z. Granate M.30," fitted with a percussion fuze of the CHZR type, is used with the 8 c.m. L. Kanone M.30.

The particulars of the design are as follows:—

Calibre, 80 m.m. (3·15 inches).
Weight (complete), 8·03 Kg. (17·66 lb.).
C.R.H. (approx.), 5.
Base streamlined $6\frac{1}{4}$ degrees.
Driving Band, copper.
H.E. Capacity, 8·9 per cent.
Bursting Charge, T.N.T. (Cast), 0·714 Kg. (1·575 lb.).
Exploders, pressed T.N.T.

An interesting feature of the design is the curvature of the streamlined base. This, however, according to our experience, gives no material advantage over a straight taper.

GERMAN COLLAPSIBLE BOMB CONTAINER
(Fig. 33)

The purpose of the container is the carriage of thirty-six 1 Kg. incendiary bombs. The whole container is released from the aircraft and falls for about five seconds before collapsing and releasing the bombs, the collapse being effected by a clockwork mechanism set in motion at the moment of release from the aircraft, by an electro-magnetic unit.

The electro-magnetic unit is energized from the Rheinmetall fuzing system.

General description

The container consists of three side pieces, A, B and C, and a separate end piece, D. A second end piece, E, is hinged to A. The sides are of aluminium of about 14 S.W.G. and the end pieces appear to be a light, strong alloy. Felt washers are glued to the insides of the end pieces. The piece B has the following attachments:—

1. A lug for attachment to the bomb carrier. This lug could be removed and screwed into the end of the central rod, enabling the container to be hung on the light series carrier of the E.S.A.C. 250 vertical bomb cell, four containers being carried in each cell.

2. A connector which fits the Rheinmetall fuzing socket of the bomb carrier.

3. A release unit. This contains a spring-driven vibrating pallet, which is held fast by a lever. The lever is released from the pallet by a small electro-magnet, thus

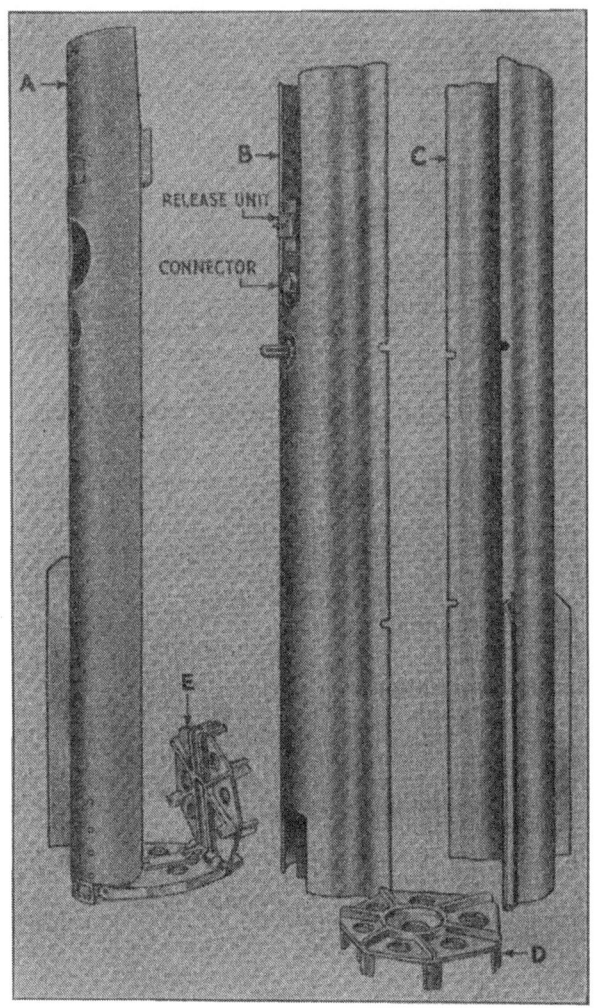

FIG. 33.

allowing the clockwork mechanism to function and rotate a spindle. The spindle holds a small steel strip in place, and after five seconds from the starting of the clockwork mechanism, turns to such a position that the strip can be pulled out. The release magnet coil is marked 24V, though it does not operate on less than 36 volts. Its resistance is 2,000 ohms, and it would appear to operate on the 150-240-volt charging current without danger of burning out in the very small time during which the current flows. Marks on the surface of the piece " A " examined correspond with the crutch position of the tier stowage carriers of the JU.88 aircraft.

Stowage of bombs

It appears that the bombs would be stowed in three tiers of twelve with partitions between each tier. The only known German bomb suited to the dimensions of the container is the 1 Kg. incendiary. The container with 36 of these bombs would weigh about 42 Kg., which is well within the limit of 50 Kg. for the carrier.

Loading and release of container

The sequence of operations would appear to be as follows:—

Loading of bombs in containers

The bombs are placed on the side piece C with the partitions between them, and the piece B is placed over them. The end pieces D and E and the central rod are then fitted, and the piece A is folded over B. The safety pin is next fitted in the end piece D, securing it to A, and the release strip and pin are inserted in the release unit, which is then cocked by turning back its spindle (thus winding the clockwork mechanism and locking the release strip in place).

Loading on aircraft

The container is attached to the carrier hook, crutched up, and the fuzing plug is inserted in the connector. The safety pin is then withdrawn.

Release

This is effected in the usual way, the collapse of the container is initiated by the charging of the Rheinmetall fuze gear. The container falls for five seconds as a complete unit. At the end of this time the clockwork mechanism allows the release strip and pin to be withdrawn by the pressure of the slipstream tending to force the pieces A and B apart. The hinged piece E then opens and withdraws the fork from the end of the rod, allowing the pieces D and E to fall away and release the sides B and C.

ITALIAN " SPEZZONE " TYPE ANTI-PERSONNEL AIRCRAFT H.E. BOMB (Fig. 34)

This type of bomb consists of a grey cylindrical body, specially designed for anti-personnel fragmentation, containing a bursting charge of T.N.T. and fitted with a detonator and an " always " fuze with safety device. Three varieties of the type have been discovered differing in size and in the construction of the body for fragmentation. The varieties may be summarized as follows: —

Weight and dimensions

	Type " A "	Type " B "	Type " C "
Weight ..	1¾ kgm.	1¾ kgm.	4 kgm. (Suspect).
Diameter ..	2·75 ins.	2·75 ins.	3 ins.
Overall length ..	6 ins.	6 ins.	6 ins.

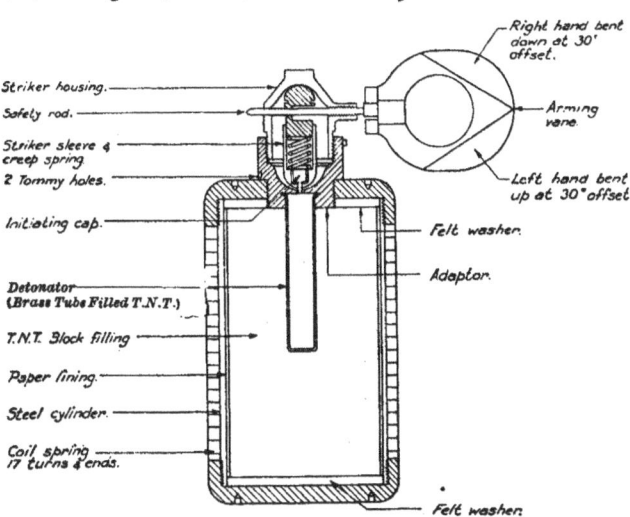

FIG. 34.

Construction of body for fragmentation

Type " A."—Thin steel cylinder with substantial ends. Cylinder is surrounded by coil spring which breaks into pieces 1 inch to 2 inches long × 0·2 inch × 0·18 inch.

Type " B."—Thin steel cylinder with light ends enclosing around an inside case coil spring similar to that in type " A."

Type " C."—Double walled steel cylinder. Space between walls filled with fragments embedded in concrete.

The upper end piece of the cylindrical body, in each case, is fitted with a fuze adapter and the lower end piece is designed to permit of block filling.

The fuze adapter is threaded to receive the fuze and has a saucer-shaped interior to assist the functioning of the "always" fuze. On the underside it is prepared to receive the detonator.

The bursting charge consists of a block of T.N.T. recessed to receive the detonator. The block is protected by felt washers at each end and round the sides by paper packing.

The detonator consists of a brass tube 2·2 inches long with a diameter of 0·35 inches. The tube, closed at the lower end and filled T.N.T., is secured into the underside of the fuze adapter.

The fuze consists of a striker housing, striker pellet, striker sleeve, creep spring and safety rod. The striker housing is of brass and is screw-threaded externally for insertion in the fuze adapter. The upper portion of the interior is saucer-shaped, to assist in the functioning of the striker pellet. The housing is also prepared to receive the safety rod. The striker pellet is cylindrical in shape with the striker formed on the inner end and the outer end rounded to bear against the saucer-shaped portion of the housing. A transverse hole passing through the pellet receives the safety rod. The striker sleeve carries the detonator in its lower end, which is rounded where it bears against the saucer-shaped interior of the fuze adapter. Within the sleeve, interposed between the detonator and the striker, is the creep spring. The safety rods consists of a rod carrying a screw thread near its outer end to engage a similar thread in the striker housing. The outer end of the rod carries an arming vane.

Action

On release the arming vane is rotated by the stream of air during the fall of the bomb. This rotation unscrews the safety rod from the housing, allowing it to fall clear, thus leaving the striker and the detonator held apart by the creep spring only. On impact, if the bomb falls on its—

(a) Head—the striker sleeve overcomes the weak creep spring by its inertia and draws its detonator on to the striker. The flash from the fuze detonator passes directly into the detonator immediately below the fuze adapter, which in turn detonates the bursting charge and fragments the bomb.

(b) Base—the inertia of the striker pellet drives the striker on to the detonator.

(c) Side—the inertia of the striker pellet working on the saucer-shaped interior of the striker housing and of the striker sleeve working on the corresponding portion of

the fuze adapter forces the pellet and sleeve together as they slide down the saucer-shaped surfaces, thus bringing the detonator and striker together.

Unfuzing

(a) The bomb is in a safe condition if the safety rod is in position. If the safety rod is missing, hold the bomb in the fuze down position for carrying or unfuzing.

(b) Unscrew and remove the fuze adapter, thus removing the detonator from the recess in the T.N.T. bursting charge. Alternatively, this may be accomplished by removing the screwed base piece and withdrawing the T.N.T. block.

(c) Unscrew the detonator from the fuze adapter.

(d) Unscrew the striker housing from the fuze adapter.

(e) Remove the striker unit complete with its detonator.

ITALIAN FUZE FOR DESTRUCTION OF ABANDONED AIRCRAFT (Fig. 35)

The fuze is designed to be used with 1 Kg. and 2 Kg. incendiary bombs or the 2 Kg. Spezzoni H.E. bomb to enable Italian.

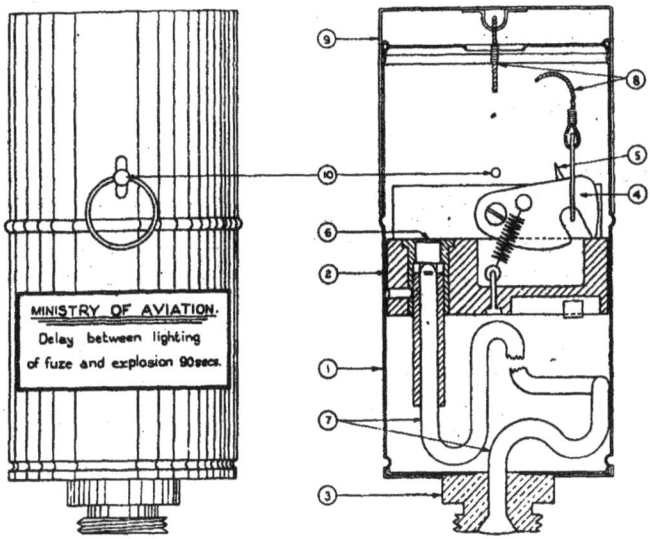

FIG. 35.

crews to destroy their aircraft when forced down in territory under British control. The bomb, fitted with the fuze for this purpose, is securely attached near the petrol tank or other vulnerable part of the aircraft and should be removed complete with fuze when found. The fuze can be removed by unscrewing the standard fuze adapter of the bomb, complete with the fuze. This removes the main detonator.

The fuze consists of a cylinder (1) (dimensions not yet available) divided into halves by a diaphragm (2) and fitted with an adapter at the base (3) for insertion in the standard fuze adapter of the above-mentioned bombs. Pivoted in the centre of the upper side of the diaphragm is an arm (4) carrying a striker (5). The arm is cut away on the underside for the loose attachment of the lanyard and prepared to receive one end of a spring, the other end of which is attached to the diaphragm. The diaphragm is also bored and fitted with an initiator (6) attached to a length of fuze (7) which is housed in the lower half of the cylinder. The length of fuze is led to the adapter at the base and gives a delay of 90 seconds. The lanyard (8), 23 feet in length, is housed in the upper compartment of the cylinder and is led through a central hole in the closing disc for attachment to the cover.

The cylindrical cover (9) is a close fit over the upper half of the fuze body and is fitted with a loop for the attachment of the outer end of the lanyard. A hole (10) is provided in one side of the cover, which is used in conjunction with a similar hole in the fuze body for the insertion of a safety pin. The presence of the safety pin prevents the turning of the striker arm on its axis.

Action

After withdrawing the safety pin, the fuze cover is removed and carried away as far as the attached lanyard will permit, *i.e.*, 23 feet. On pulling the lanyard further the striker arm is rotated on its axis, thereby putting further tension on the spring. When the striker arm has passed the central position the striker is driven on to the initiator by the spring and the lanyard becomes detached. The length of fuze is ignited by the initiator and at the end of 90 seconds the bomb explodes.

ITALIAN 12·7 m.m. INCENDIARY TRACER
(Fig. 36)

The cartridge case is of brass and of the semi-rimless type.

The boat-tailed bullet consists of a steel body (a) heavily coated on the exterior with gilding metal (b) and with a tracer cavity (c) formed in the base. The tracer composition gives a white trace up to a distance of approximately 150 yards, the trace then changes to red. The incendiary filling (d) consists of aluminium or potassium chlorate with a small layer of

penthrite/wax on top (e). Situated above the incendiary filling is the initiator arrangement. The brass fuze is similar in action to the German A.Z.5045, although the centrifugal segments with spring coil are housed in the bottom half, instead of in the nose-cap, and the striker incorporates the hammer. The body of the bullet is painted blue.

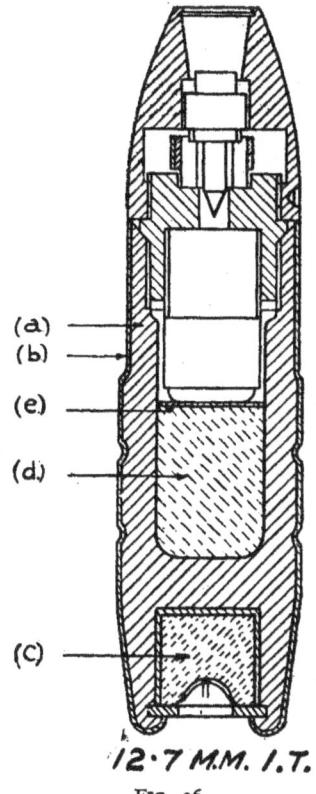

12·7 M.M. I.T.

Fig. 36.

ITALIAN 12·7 m.m., A.P., INCENDIARY TRACER
(Fig. 37)

The cartridge case is of brass and of the semi-rimless type. The boat-tailed bullet consists of a cupro-nickel envelope (a), copper sleeve, also forming the nose of the bullet (b), steel core (c) with lead band (d) pressed around its base into canne-

lures, a lead base plug (e) and tracer tube (f), transparent closing disc (g) and brass washer (h), and split steel support piece (j) in the nose which possibly acts as a forming anvil. The nose (of the copper sleeve) is filled with three compositions, first in the tip and intimately pressed into the split anvil, a mixture of aluminium and lead oxide. Next a mixture of magnesium, potassium chlorate and antimony sulphide, and

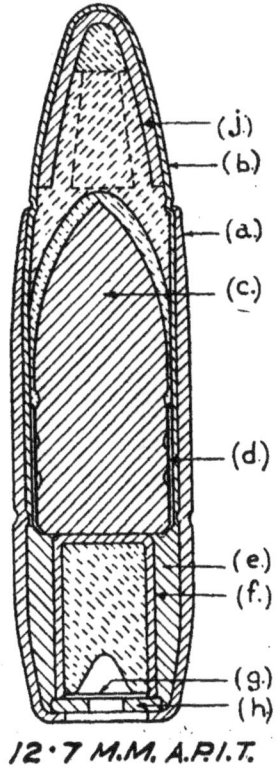

12·7 M.M. A.P.I.T.

FIG. 37.

round the nose of the core, reaching back to the lead band at the base of the core, a mixture of aluminium, magnesium and lead oxide. The central composition is apparently intended to act as an igniter for the other two incendiary compositions. The tracer composition gives a bright red trace. The tip of the bullet is painted white to the joint with the cupro-nickel envelope.

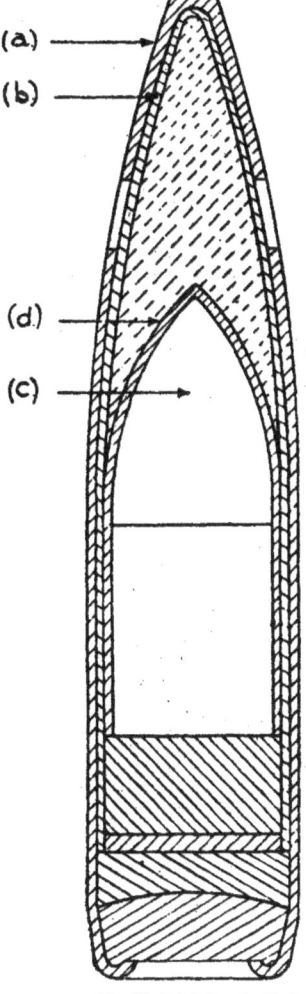

7·7 M.M. A.P.I.
Fig. 38.

ITALIAN 7·7 m.m., A.P., INCENDIARY (BLUE TIPPED) (Fig. 38)

In general appearance this round is similar to the normal British ·303 inch type. The nose of the bullet envelope has four holes and 0·5 inch of the tip is painted blue. The brass cartridge case is fitted with a copper cap. The cap is not ringed in.

The bullet consists of an envelope (a) of steel coated with cupro-nickel (S.M.1) or cupro-nickel (B.P.D.), copper sleeve (b) and a steel core (c) in a lead sheath (d). The nose portion of the copper sleeve is filled phosphorus and the envelope is pierced with four holes on the ogive, so that at these points the phosphorus is protected only by the copper sleeve. This presumably permits collapse and distribution of the phosphorus on impact with a sufficiently rigid material. The methods of closing the base of the bullet vary, but a typical example is shown. In each method there is a solder seal.

Base Markings:—   B.P.D.   S.M.1.
                       37   936

ITALIAN 7·7 m.m., A.P., INCENDIARY (GREEN TIPPED) (Fig. 39)

In general appearance this round is similar to the normal British ·303 inch type. The nose of the bullet envelope has four holes and 0·5 inch of the tip painted green. The brass cartridge case is fitted with a copper cap. The cap is not ringed in.

The make-up of the bullet is almost identical with that of the blue-tipped bullet described above, but the nose of the copper sheath is filled with two compositions. In the tip is a small quantity of magnesium and potassium chlorate. Behind this and surrounding the nose of the core is aluminium and lead oxide. Presumably the first mixture acts as an igniter to the thermite type of composition behind it.

Base Markings:—   B.P.D.
                       39

ITALIAN, 4·5 Kg., MANZOLINI, H.E., AIRCRAFT BOMB (Figs. 40 and 41)

This bomb, an invention of Commandatore Manzolini, is also known as the "Thermos" bomb because of its resemblance in appearance to a thermos flask. It is an H.E. bomb fitted with a fuze which becomes fully armed after impact and subsequently initiates the detonation of the bomb when disturbed by handling, etc.

The body of the bomb is constructed of $\frac{3}{8}$ inch seamless mild steel tube $2\frac{1}{2}$ inches in diameter and 7 inches long, prepared

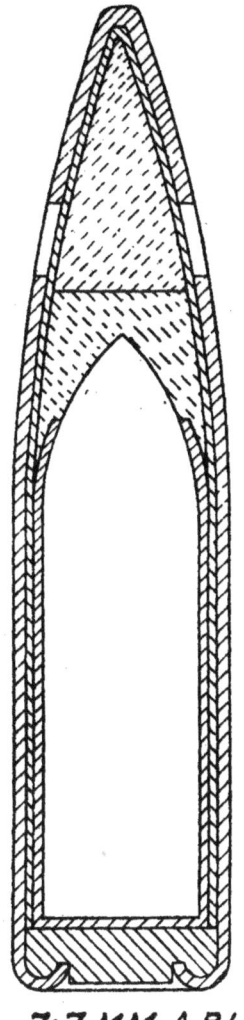

7·7 M.M. A.P.I.
Fig. 39.

at the tail end to receive the fuze and closed at the other end by a steel disc which is welded in position. The bursting charge consists of approximately 2 lb. of T.N.T. The body is painted brown and the overall length (with fuze) is 14 inches.

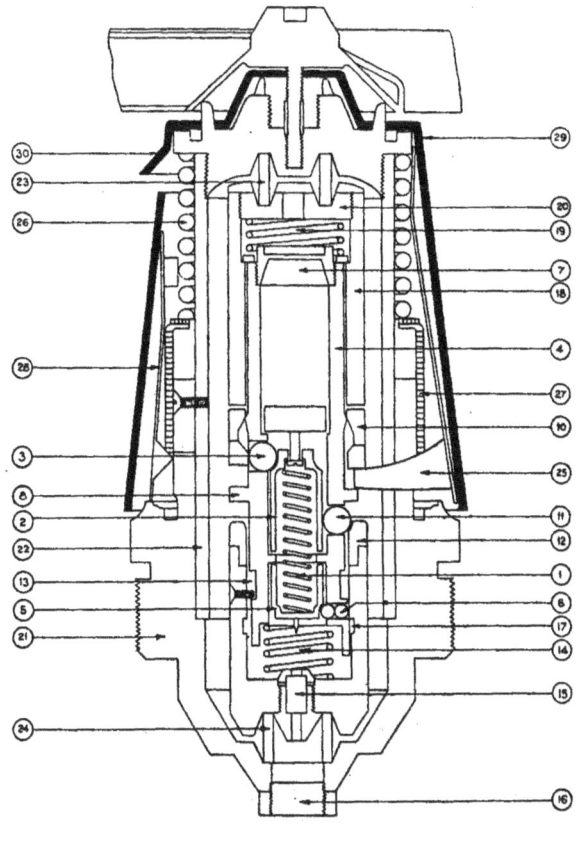

FIG. 40.

The fuze consists of a striker spring (1) held under compression between two cup-shaped containers, one fitting over each end of the spring. The cup fitted over the tail end of the spring (2) carries a piston and is prevented from rising by three balls (3), located in the walls of the liquid cylinder (4), which

bear on a shoulder formed on the cup. The cup fitted over the other end of the striker spring (5) carries the striker and is prevented from moving out of the liquid cylinder by three pairs of steel balls (6) also located in the wall of the liquid cylinder and bearing on a shoulder formed on the cup.

The liquid cylinder is of brass and is closed at the tail end by a screwed plug (7). A circumferential flange (8) is formed externally about its centre to limit its downward movement within the front cup. Above and also below this flange the cylinder is prepared to house two sets of three steel balls. The upper set prevents the piston cup from rising and the three balls are retained in position by an inertia sleeve (10). The three steel balls below the flange (11) are retained by the piston cup and the open end of the front cup (12). While these balls are in this position further movement of the liquid cylinder into the front cup is prevented. Near the front end the liquid cylinder has a circumferential groove (13) which in combination with a screw carried in the front cup limits the axial movement of the cylinder. Below this groove the cylinder is prepared to house three pairs of steel balls which support the striker cup. At the front end the cylinder is recessed to accommodate a weak spiral spring (14). Internally the cylinder is designed to form an oil chamber for the piston and a circumferential flange is formed to position the piston and striker cups.

The front cup, which receives the forward end of the liquid cylinder, is of brass and carries a small detonator (15) positioned over a flashhole leading to the larger detonator (16) which is carried in the body of the fuze and is in contact with the bursting charge. The cup has an internal circumferential groove (17) to receive the steel balls supporting the striker cap when the fuze is fired.

The tail cup (18) which fits over the tail end of the liquid cylinder is of brass and encloses a strong spiral spring (19) held under compression between the tail end of the liquid cylinder and the closing plug of the tail cup (20).

The body of the fuze (21) is screw-threaded for insertion in the bomb and is fitted with a hollow steel cylinder (22) within which the liquid cylinder assembled with its front and tail cups, etc., is held by means of two loose tubular positioning pieces, one above the tail cup (23) and the other below the front cup (24). Three holes are provided in the wall of this body cylinder, near its centre, to receive the claws (25), fitted to spring strips to support the inertia sleeve of the liquid cylinder. The body cylinder is closed at the tail end by a brass screwed plug which is prepared to receive the screw-threaded shaft of the arming vanes and is flanged to support one end of the arming spring (26). The arming sleeve (27) is a short blackened cylinder with rubber insert which fits over

the body cylinder and is positioned between the arming spring and the claws with three distance pieces (28). The distance pieces and the three spring strips carrying the claws are

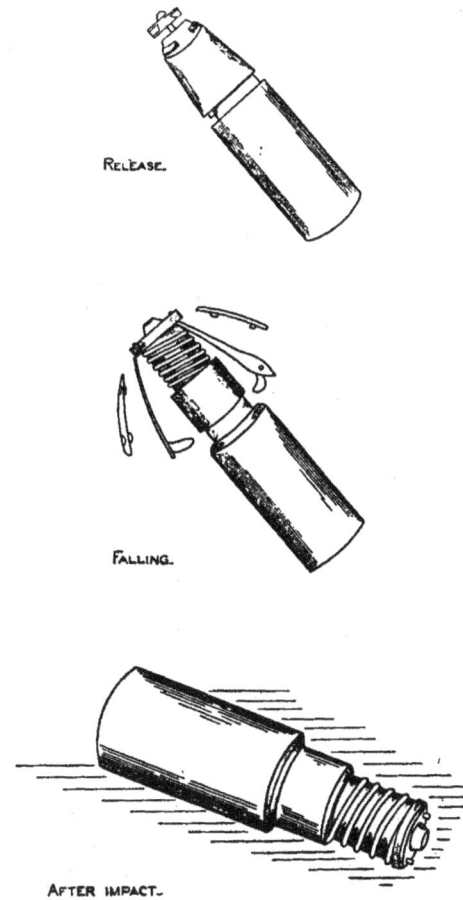

FIG. 41.

retained in the engaged position by an aluminium tail cap (29) which is secured over the tail of the fuze by the arming vanes. Projecting louvre-like fins (30) are formed in the end of the tail cap.

The arming vanes are of aluminium, one of the vanes being prepared to receive a safety pin which, when in position, prevents rotation of the vanes.

Action

Before release the safety pin is removed from the arming vanes, thus allowing the vanes to be rotated by air resistance during the descent of the bomb. The rotation of the arming vanes unscrews the shaft of the vanes from the brass cap of the body cylinder and releases the aluminium tail cap, which is then removed from the fuze by the spring strips carrying the claws and the action of the air against the fins of the tail cap. The removal of the tail cap permits the claw spring strips and the distance pieces to fall away and the arming sleeve is moved by its spring to seal the claw holes in the body cylinder. This movement of the arming sleeve assists in the withdrawal of the claws and the removal of the tail cap. The withdrawal of the claws removes the support between the inertia ring and the external flange on the liquid cylinder and leaves the ring supported by a small projecting ring formed on the cylinder.

On impact the tail cap sets down over the liquid cylinder, compressing its spring and driving the inertia ring down on to the external flange of the liquid cylinder. This movement brings the groove on the inside of the inertia sleeve in line with the steel balls in the wall of the liquid cylinder which retain the piston cup of the spring. These balls enter the groove and allow the piston to rise under the force of the spring and against the resistance offered by the oil flowing through the clearance between the piston and the cylinder wall. The upward movement of the piston removes the support of the piston cup from the steel balls located in the wall of the liquid cylinder below the flange and permits the balls to enter the cylinder, thus removing the means of preventing the liquid cylinder from further entry into the front cup. This further entry is now opposed by the weak spiral spring below the liquid cylinder and the fuze is fully armed.

On being subjected to a sudden movement or jarring action the opposition of the spring is overcome by the movement of the liquid cylinder into the front cup or that of the front cup over the liquid cylinder, according to the direction from which the disturbing force is applied. Either of these movements brings the groove inside the front cup in line with the three pairs of steel balls located near the front end of the liquid cylinder which support the striker cup. The balls enter the groove and the striker is driven by its spring on to the initiating detonator and the detonation of the bursting charge is brought about through the main detonator with which it is in contact.

From the drawings available the fuze appears to be designed

to function also on the " always " principle when the disturbing force is applied laterally. In these circumstances a sudden movement or jarring action would cause the liquid cylinder to reach the firing position within the front cup as the result of the closing movement of both of these components. The movement of the liquid cylinder would be brought about by the convex surface of the closing plug of the tail cup moving down the concave underside of the closing plug of the body cylinder and transmitting the movement to the liquid cylinder through the intervening spring. Corresponding movement of the front cup, in the opposite direction, would be brought about by the incline on the front end of the front cup moving down the corresponding incline in the fuze body. In each case the inclined sides of the grooves accommodating the tubular positioning pieces appear to facilitate these movements.

The sensitivity of the fuze is dependent on the strength of the spiral spring in the front cup. The existence of a green bomb of this type fitted with a more sensitive fuze has been reported.